The Marvelous World of 2D Materials: Unlocking a New Era in Technology

Mariyak

Contents

Chapter 1

Introduction

1.1   2D materials

The field of nanomaterials studies structures with dimensions in the nanometer range (1nm-100nm). If only one dimension of the considered material is in the order of the nanometer, the material is called 2D. Similarly, if the materials are restricted to nanometer in 2 dimensions, the structure is one-dimensional or a wire. If all dimensions are restricted to nanometer scales we consider it a 0D structure, as for example, quantum dots or fullerenes [10–12]. In general, there are more nanomaterials with one or more dimensions in nano-size. Figure [1.1a] shows the classification of nanomaterials based on restricted dimensions. 2D materials are of a broad interest for researchers because of their extraordinary properties, such as charge mobility, tunable optical properties, good mechanical strength and long spin diffusion length for spintronics devices [13]. Both 2D materials applications and the latest improvement

in exfoliation techniques have pushed the production of single layer sheets from different
bulk van der Waals materials used for building of heterostructures with desirable electronic
properties (See Figure 1.1b). Furthermore, the 2D material family includes combinations
that involve almost all elements in the periodic table. This leads to different electronic
properties, as semimetals, metals, insulators, semiconductors and insulators with different
bandgaps [14]. The 2D materials are classified based on their structure, composition and
electronic properties. In fact, 2D materials are anticipated to have a significant effect on a
large variety of applications, such as flexible electronics and high-performance sensors [12].

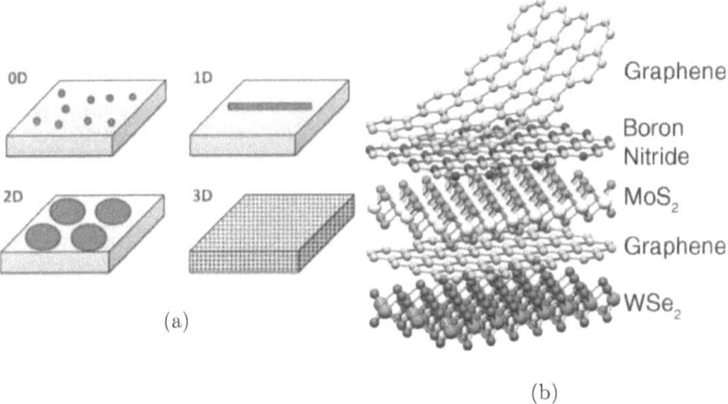

Figure 1.1: (a) Schematic of dimensionality of materials and their nanostructures [1]. (b)
Schematic of heterostructure by stacking of 2D materials [2].

1.2 Fabrication of nanomaterials

Nanostructures can be made in various ways using different techniques. Top-down and bottom-up are two approaches for synthesis of nanomaterials [15]. In the top-down approach the bulk material breaks down to increasingly smaller parts until we get the desired structure with a size less than 100nm. This process uses lithography. In the bottom-up approach, nanostructures form atom by atom and layer by layer [15, 16].

1.2.1 Bottom-up

The bottom-up approach is a technique to build up nanomaterials from the bottom, which means starting with the atomic ingredients and assembling them, atom by atom, molecule by molecule or cluster by cluster and so on.

1.2.2 Top-down

The top-down approach indicates successive cutting of bulk material to get 2D-sized materials. The nanomaterial in the bulk form becomes smaller and smaller, (See Figure [1.2]). In other words, we start with the bulk material and make it thinner.

1.2.3 Comparing fabrication methods

Regarding technical aspects, usually, the bottom-up method is chemically performed (synthesis), while the top-down method is a nano-fabrication procedure. This does not mean that the top-down approach may not involve chemical procedures, but we primarily start with bulk (or macro) structures and decrease their sizes. By contrast, in the bottom-up method, we start with a homogeneous solution or gas and build up the nano-structure or nano-layer, nano-wire, etc [15,16]. For example, milling is a top-down approach for generating nanoparticles while colloidal dispersion is a bottom-up method of synthesis of nanomaterials. There are some advantages and disadvantages for each method. The main challenge in the top-down method is the creation of defects at the surface as the structure details become smaller, which leads to difficulties in device design and fabrication. Besides the imperfections produced by a top-down approach, the method will continue to play an essential role in the synthesis of nanomaterials [17].

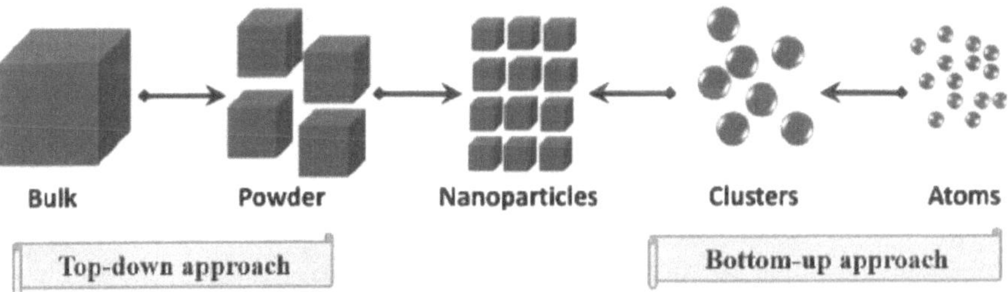

Figure 1.2: Top-down and bottom-up approaches for the synthesis of nanoparticles [3].

1.3 Examples of 2D materials

1.3.1 Graphene

A single layer of Graphene is the first 2D material isolated experimentally (2004) by exfoliation technique [18]. Graphene was predicted and theoretically studied many years before that [19]. Since 2004, many researchers have discovered more properties of these atomically thin materials thus opening a new and interesting world of useful applications. Graphene is a plane of carbon atoms, one atom thick, arranged in a honeycomb structure (See Figure [1.3a]). In this sense, graphite is a collection of stacked graphene layers kept together by the Van der Waals force, Figure [1.3b]. So, graphene is a 2D material with atomic thickness; the decrease in dimensionality gives rise to unique physical and chemical properties which make it substantially different from 3D structures. Among 2D materials, graphene is perhaps the most well-studied, with well-known properties that make it special and attractive, both from a fundamental point of view and its potential applications. Graphene has two atoms per unit cell and its electronic band structure is described by conical valence and conduction bands touching each other in two inequivalent points, K and K', in the Brillouin zone (BZ), which are called Dirac points. Its low energy properties are dominated by out-of-plane p_z orbitals [20]. It is a gapless material, and an energy gap is essential to switch between metallic and insulating states. This feature makes graphene inappropriate for some electronic devices. For example, the bandgap in the visible or infrared range of the spectrum is required for the solar cell and telecommunication applications. As a result, significant efforts have been made to understand all possible 2D semiconducting crystals [10, 21].

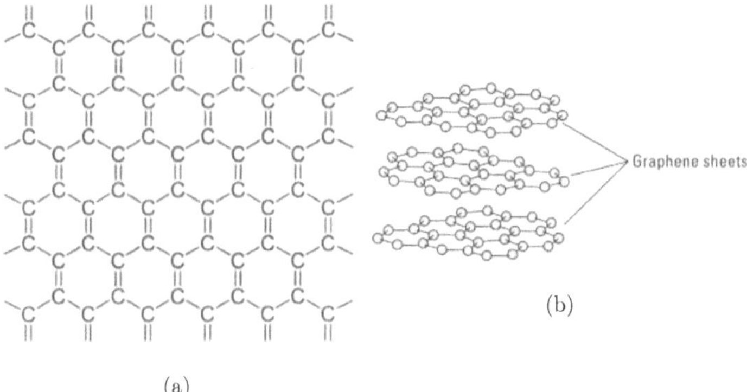

Figure 1.3: (a) The Graphene layer consists of a single layer of carbon atoms arranged in a honeycomb structure [4]. (b) Graphite as the stack of graphene sheets [5].

1.3.2 Transition metal dichalcogenides (TMDs)

TMDs' structures form an extensive and rich family of crystals, which includes members representing a diverse set of lattice structures with distinct physical properties. The semiconductor compounds that have been widely studied are of the type MX_2 (M = Mo, W; X = S, Se, Te). Here, M is the transition metal element and X represents the chalcogen. The transition metal elements can be from the group IV (Ti, Zr, etc.), V (V, Nb, etc.) or VI (Mo, W, etc.) Figure [1.4a]. They are composed in the bulk of the two-dimensional X–M–X layers stacked on top of each other and coupled by weak Van der Waals forces. MX_2 hexagonal layers (graphene-like) represent a single sandwich structure where the M metal atoms are arranged in a triangular lattice, each M atom bonds to six chalcogen atoms covalently, three of X atoms in the top and three in the bottom. From a side view the MX_2 layer consists of three layers X, M, X in this order [22, 23]. The electronic band structure of these materials strongly depends on the number of layers. A single layer of TMDC has direct bandgap of approximately 1.9 eV placed at the K and K' points of the hexagonal Brillouin zone. On the other hand, a multilayer of TMDCs has an indirect bandgap of about 1.3 eV with the highest valence band at Γ point and the lowest conduction band state at the middle point along the $\Gamma - K$ path [21]. Furthermore, TMDs have a strong spin-orbit coupling, which together with the lack of inversion symmetry results in a splitting of the valence band of about 140 meV (for Mo compounds) and about 400 meV (for W compounds) [24]. The conduction band is also split by tens of meV. Due to the presence of bandgap in 2D TMDC semiconductors and its magnetic properties, these materials provide the properties that graphene cannot [11, 25].

Molybdenum disulfide (MoS_2) is a member of transition metal dichalcogenide (TMDC) family [26, 27]. The MoS_2 crystal structure has the form of a hexagonal lattice of Mo atoms on one sublattice and a pair of S atoms in the other one. As discussed before, from a side point of view, this arrangement appears as three shells, or planes. Atoms are shown in Figure [1.4b]. These three planes are connected to each other by strong covalent bonds between the Mo and S atoms; however, in the bulk material, there is a weak Van der Waals force holding the MoS_2 layers together. Because of this distinction MoS_2 layer is considered a 2D material. MoS_2 like most semiconducting TMDCs is described by a thickness-dependent bandgap that has been predicted theoretically [28] and confirmed experimentally [29, 30]. The bulk form of MoS_2 has an indirect bandgap of about 1.2 eV. On the other hand, in its monolayer form, the bandgap rises to about 1.8 eV [9] because of quantum confinement effects [27]. A single layer of MoS_2 has different properties compared to its bulk [31].

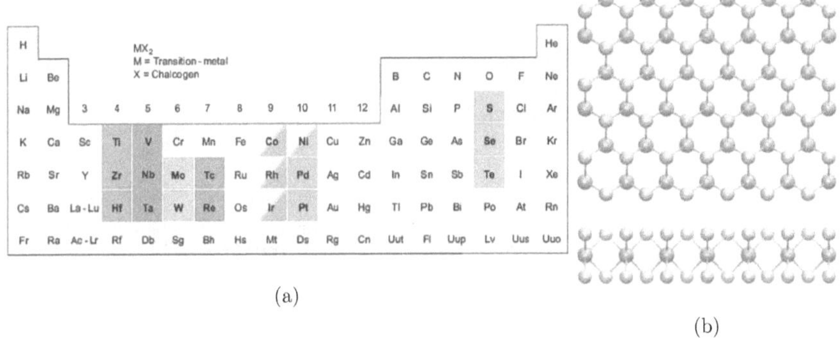

(a)

(b)

Figure 1.4: (a) The list of elements found in TMD materials consists of 16 transition metals [6] (M) and three chalcogens (X) (b) Top and side view of MoS_2 [7]

1.3.3 Hexagonal boron nitride (h-BN)

The h-BN is an insulator material with two different atoms B (Boron) and N (Nitrogen) arranged in different sublattices of hexagonal lattice. It has a direct bandgap of about 4.5 eV placed at K and K' points of the Brillouin zone [21]. Furthermore, due to its insulator properties h-BN is an excellent substrate for supporting and encapsulating some materials such as graphene, black phosphorus, etc. For example, this material is widely used in 2D materials research as a dielectric substrate instead of SiO_2. Graphene can improve device performance, due to its large electronic mobility and very low concentration of trapped charges. Moreover, graphene on h-BN is a perfect platform to study Moiré type effects on the electronic and optical properties due to its small lattice mismatch with graphene, which is around 1.8% [21]. In addition, h-BN is an essential building block in the so-called Van der

8

Waals heterostructures. These structures are characterized by the stacking of different 2D crystals bonded by Van der Waals forces. The properties of such structures are determined by the choice of the stacked 2D materials [8].

1.4 Properties of two dimensional materials

Two-dimensional structures are fascinating new materials, attracting the interest of many researchers. They look for a way to explore its unique properties, very distinct from its 3D counterparts. As mentioned in the introduction, limiting the leading dimensions of a material will directly impact its material properties. Due to electron confinement and the absence of interlayer interactions, not only the optical and electronic properties are affected, but also mechanical and chemical aspects. The interlayer interactions are usually quite weak, playing some impact on the shape of the band structure. Usually, this is because of geometry effects and due to the high, even infinite, surface-bulk ratio in the thinnest materials [12].

1.4.1 Size effects

In nanostructures, size effects become more and more important while decreasing the critical dimension, in other words, when the scale length of the physical phenomenon (coherent length, screening length, for example) become comparable to the characteristic size (length, thickness, diameter, and others) of the nanostructure. The properties of nanometric systems are described by a specific "length scale" generally in the order of the nanometer. Physically, if the size and shape of the materials change in such a way that takes its dimensions below the specific length scale, its properties will change. Thus, size effects introduce special and attractive aspects for research in nanomaterials. Furthermore, size effects relate to thermodynamic, electronic, chemical, electromagnetic, and structural features of these structures [32]. Classical physics laws cannot describe the observed properties of materials in the nanoscale, to throw some light on this subject it is necessary to explore its quantum properties.

1.4.2 Quantum Confinement

We can say that nanomaterials are closer in size to single atoms and molecules than to bulk materials. Thus, quantum mechanics is necessary to understand their behavior. Essentially, quantum mechanics is the part of physics that models, or studies, the motion and energy of atoms and electrons. The most remarkable quantum effects together with other physical properties are as follows. First, nanomaterials are very small structures (< 100 nm), so their mass is very small and gravitational forces become negligible. Therefore, electromagnetic forces are dominant in determining the dynamics of atoms and molecules. Furthermore, according to the wave-particle duality of the matter, matter that has very small mass, like the electrons, has a pronounced wave-like nature. Consequently, electrons exhibit wave behavior and their dynamics relate to movement of a probability wave function. Moreover, classically an object can go over a potential barrier only if it has appropriate energy to "jump" over it. Thus, if the object has lower energy than that required to jump over this energy barrier, or obstacle, the probability of detecting the object on the other side of the barrier is zero. On the other hand, in quantum mechanics a particle with energy lower than the energy required to jump the barrier has a finite probability of tunnel to the other side of the barrier [33].

1.4.3   Van der Waals (vdW) materials

The amazing property of these materials is that they can be joined together to build new materials. These 2D layered materials have covalently-bonded atomic layers stacked and joined together by the weak Van der Waals (VdW) interaction; that is a rich environment for study and engineering of 2D materials [34]. As powerful covalent bonds cause the stability in-plane for 2D crystals, these materials are called Van der Waals heterostructures because the atomically thin layers are not mixed through a chemical reaction but rather connected to each other by a weak so-called Van der Waals interaction, in the same way one can stick a tape to a surface [35]. Generally, the bonds in 2D materials that interact between atoms in the plane are strong because all atoms and molecules attract each other via the ubiquitous Van der Waals (vdW) forces. It is still not totally clear how all of these new excellent thin materials stack to each other just like Lego blocks (See Figure 1.5). Based on the ability to stack any number of atomically thin layers, Van der Waals heterostructures have a large potential to generate many metamaterials and new devices [36]. There are many possible combinations allowing to further reach new uninvestigated optoelectronic device functionalities or remarkable material properties previously inaccessible in traditional 3D material science. In addition to the coupling of various 2D atomic layers by vdW force, the passivated, dangling-bond-free surface of a 2D crystal can bond with materials with different dimensionalities. The creation of mixed-dimensional vdW heterostructures could be done by hybridizing 2D crystals, mostly graphene, with 0D quantum dots or nanoparticles, 1D nanostructures, nanowires or carbon nanotubes, or 3D bulk materials [8]. In this regard, an enormous number of 2D atomic crystals could be mechanically peeled from bulk single crystals. Moreover, there are no physical bonds among the layers, making it easy to mechanically stack arbitrary 2D materials onto each other similar to gathering atomic-scale Lego blocks.

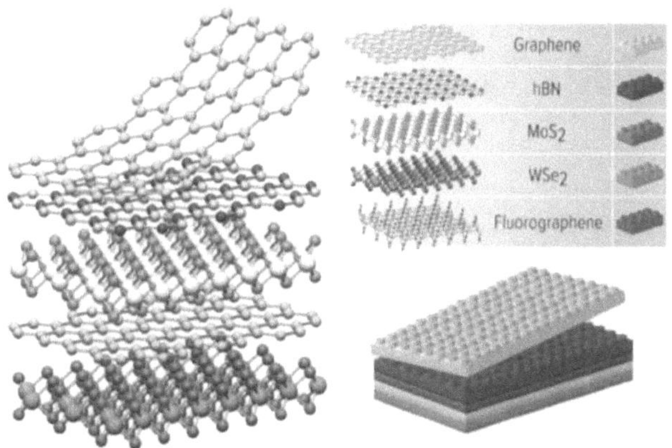

Figure 1.5: Building Van der Waals heterostructures [8].

1.4.4 Electronic and optical properties

There is a big number of 2D materials, including the graphene family, metal chalcogenides and metal oxides in nature [37]. These materials exhibit a wide spectrum of electrical and optical properties, as shown in Figure [1.6]. For the 2D semiconductors, an assortment of transition metal dichalcogenides (TMDCs) involving IV groups like (Mo and W) dichalcogenides, are available. The semiconducting TMDCs have a medium size bandgap, contrasting with graphene that has no bandgap. They have been used in electronic devices including field- effect transistors (FETs) and memories [38]. Furthermore, they are promising optical

12

materials as their energy spectrum is in the range of visible to near-infrared wavelength [39].
These atomically thin 2D semiconductors also have special optical properties including the
indirect- to-direct bandgap transition in the one-layer regime [40], good spin-valley coupling [41], strong light–matter interactions [42], and large exciton binding energy. Recently,
black phosphorous has emerged as a new 2D semiconductor besides semiconducting TMDCs.
Specifically, it has a tunable direct bandgap, mostly changing from 0.3 eV to 2.0 eV with
decreasing thickness, or number of layers. Hexagonal boron nitride (h-BN) is a large bandgap
insulator with a bandgap of about 5 eV which has drawn a lot of attention in recent times.
As the h-BN has a quite flat and smooth surface, with a shortage of relaxing bonds, it has
been successfully applied as an ideal substrate for other 2D conducting channels, for instance,
graphene and TMDCs [43].

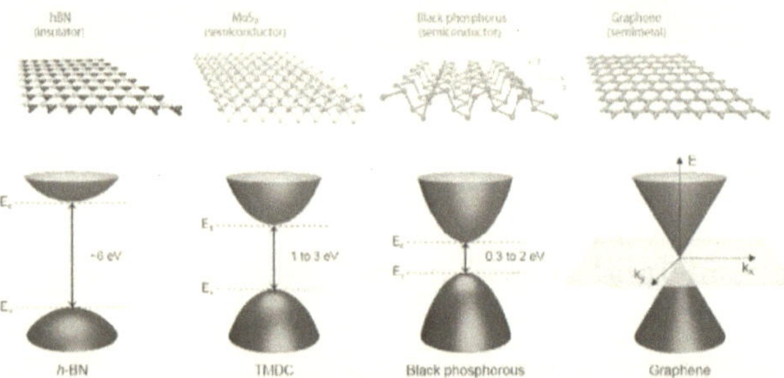

Figure 1.6: Energy spectrum and atomic crystal structures for different two-dimensional
(2D) materials. From left to right: boron nitride (h-BN), transition metal dichalcogenides
(TMDCs), black phosphorous, and graphene.

Indeed, due to a stunning variety of different ranges of electrical, chemical, optical
and mechanical properties of 2D materials, the pivotal discovery about these materials is that

they can be joined together to build new materials. Researchers found that by superposing two atomically thin graphene-like materials on top of each other it is possible to achieve different properties for the new material and even attain new properties. These new properties can be explored in new materials and nano-devices. The properties of these new materials can be precisely controlled by twisting the two stacked atomic layers, paving the way for exploring this unique class of degree of freedom in nano-devices of combined materials in future technologies. The next generation of nanoelectronics can make formidable use of 2D materials given their unique interlayer coupling and optoelectronic properties, allowing them to create high-performance state-of-the-art nanostructures tailored to a particular purpose.

Chapter 2

Methodology

2.1 Introduction

To access the physical properties of solids, we should consider quantum mechanics of many electrons and ions. Including all interacting particles for an adequate electronic description makes this system difficult to investigate. This difficulty arises because there is a large number of electrons and ions to consider. Therefore, some simplification needs to be done. In this work, we will make use of the well-established density functional theory to study 2D materials [44–46].

2.2 Density Functional Theory

2.2.1 Many-body Schrödinger equation

In this part, before introducing Density Functional Theory (DFT), we will review fundamental concepts, necessary for a correct description of the electronic structure. The electronic properties of a solid with N electrons are described by a many-body Schrödinger equation [47, 48]:

$$H\psi(r_1, r_2, r_3, ...r_N) = E\psi(r_1, r_2, r_3, ...r_N). \tag{2.1}$$

Here, H is the Hamiltonian operator and ψ is the wavefunction, or eigenvector. Each ψ is associated with allowed value of energy, or eigenvalue, given by E. The Hamiltonian is composed of terms related to electrons (e), nuclei (n), and the interaction between them (en), as given by:

$$H = T_e + T_n + V_{ee} + V_{en} + V_{nn}. \tag{2.2}$$

Here T_e and T_n are kinetic energy for electrons and nuclei, respectively, and the other three terms define the potential energy of electrons in the presence of other electrons (V_{ee}), electrons in the presence of nuclei (V_{en}), and nuclei in the presence of other nuclei (V_{nn}). The way to find the ground state ψ_{ES} of a system like this is to solve the many-body Schrödinger equation that describes electrons and nuclei [49]. The Hamiltonian operator, as presented in Eq.[2.2], can be explicitly written as:

$$H = \sum_{i=1}^{N_e} \frac{p_i^2}{2m_i} + \sum_{\alpha=1}^{M} \frac{p_\alpha^2}{2\mu} + \frac{1}{2}\sum_{i=1}^{N_e}\sum_{j\neq i}^{N_e} \frac{e^2}{|\vec{r_i} - \vec{r_j}|} - \sum_{i=1}^{N_e}\sum_{\alpha=1}^{M} \frac{e^2 Z_\alpha}{|\vec{r_i} - \vec{R_\alpha}|} + \frac{1}{2}\sum_{\alpha=1}^{M}\sum_{\beta\neq\alpha}^{M} \frac{Z_\alpha Z_\beta e^2}{|\vec{R_\alpha} - \vec{R_\beta}|} \tag{2.3}$$

Here α and β indexes run over M nuclei and i and j over N_e electrons. The three parts on the right (the double summations) represent the potential energy of the system. These parts describe the Coulomb interaction between the electron-nuclei attraction, and electron-electron and nuclei-nuclei repulsions. In basic quantum mechanics scenarios, one usually considers only single electron states. In a solid, we are dealing with many particles and this fact makes things extremely complicated. To simplify the problem, we consider the Born-Oppenheimer approximation. Due to the large differences between nuclei and electron masses, we decouple the dynamics of these two distinct particles [50]. Thus, we can write our Hamiltonian only in terms of electrons in the presence of frozen position of nuclei:

$$H = \sum_{i=1}^{N_e} -\frac{h^2}{2m_e} \nabla_e^2 + \frac{1}{2} \sum_{i=1}^{N_e} \sum_{j \neq i}^{N_e} \frac{e^2}{|\vec{r_i} - \vec{r_j}|} - \sum_{i=1}^{N_e} \sum_{\alpha=1}^{M} \frac{e^2 Z_\alpha}{|\vec{r_i} - \vec{R_\alpha}|} \tag{2.4}$$

This equation [2.4] represents the desirable many-body electronic dynamics; however, it is still cumbersome to solve. For example, if we take the benzene structure, it consists of a system with 42 electrons. In this case, for three spatial coordinates, we have to solve the Schrödinger equation as a 126-dimensional problem which is a hard task to fulfill. So, we can say that solving the many-body Schrödinger equation for a given material is difficult. From here on, we describe the approach of Density Functional Theory. This approach reduces the problem dimension from 3ND to 3D spatial problem, where N is the number of particles. Density Functional Theory (DFT) deals with Schrödinger equation in terms of the electronic density.

2.2.2 Density Functional Theory (DFT)

The Density Functional Theory (DFT) achieved excellent results in describing properties of a large variety of systems: from atoms and molecules to solids and surfaces. In the DFT approach, we approximate solutions for the Schrödinger equation for the many interacting particles by solutions of noninteracting Kohn-Sham particles with correct electronic density [10]. The Density Functional Theory (DFT) is based on two fundamental theorems by Kohn and Hohenberg [45].

2.2.3 The Hohenberg-Kohn theorems

The First Theorem

"For any system of interacting particles in an external potential $V_{ext}(r)$, the density is uniquely determined; in other words, the external potential is a unique functional of the density" [48]. This theorem states that the electronic density is all that we need to describe the ground state.

$$E = E[n(\vec{r})] \tag{2.5}$$

The Second Theorem

"A universal functional for the energy E[n] can be defined in terms of the density. The exact ground state is the global minimum value of this functional." [51]. This theorem describes how the ground state energy can be defined by electronic density which minimizes the energy functional. In other words, the ground state energy is minimized by the correct ground state electronic density, which means that one should look through different electronic densities until the lowest energy is found.

$$E[n(\vec{r})] > E_0[n_0(\vec{r})] \tag{2.6}$$

2.2.4   Kohn-Sham equations

The fundamental idea of the Kohn–Sham scheme is to replace the interacting many-particle problem by a system of non-interacting particles with the same ground-state density $n(r)$ as the original many-particle system [10, 46]. This means that, for an interacting system, the ground state density is the same as the density of non-interacting system of Kohn–Sham particles.

In DFT, the total electronic energy E is a functional of the electronic density, written as:

$$E[n(\vec{r})] = T[n(\vec{r})] + \int dr n(\vec{r}) v_{ext}(\vec{r}) + E_H[n(\vec{r})] + E_{XC}(\vec{r}). \tag{2.7}$$

In this expression, the first term is the Kohn-Sham kinetic energy $T[n(r)]$ for the non- interacting system, $n(r)$ is the electronic density, and $v_{ext}(r)$ is the external Coulomb potential. The two last terms on the right-hand side of equation [2.7] describe electron-electron interactions. This functional can be divided into two main parts: a) A known part which is given by $E[n(r)] = T[n(r)] + \int dr n(r) v_{ext}(r) + E_H[n(r)]$ where $E_H[n(\vec{r})]$ is the Hartree energy. and b) the unknown part ($E_{XC}(r)$) called exchange-correlation term. The challenge in this approach is the $E_{XC}(r)$ part, since it is unknown [10]. Nevertheless, there are some well-established approximations to E_{XC}. The coulomb energy coming from the interaction between two charge densities is as expected:

$$E_H = \frac{1}{2} \int \frac{n(\vec{r})n(\vec{r'})}{|\vec{r} - \vec{r'}|} dr \, dr' \tag{2.8}$$

This expression is called the Hartree energy. The exchange-correlation energy (E_{XC}) describes effects of the exchange and correlation of electrons. This effect, due to the exclusion principle, is present in the quantum mechanics of many-body systems. Therefore, one can say that the effects of many-particles are all included in E_{XC}. Finding the E_{XC} term is difficult, but there are some good approximations for it.

Now, we can write the Kohn-Sham single-particle Hamiltonian as the following expression [52]:

$$H_{KS} = -\frac{1}{2}\nabla^2 + V_I(\vec{r}) + V_H[n(\vec{r})] + V_{XC}[n(\vec{r})] \tag{2.9}$$

Here,

$$V_I(\vec{r}) = \sum_{i=1}^{M} -\frac{e^2 Z}{|\vec{r} - \vec{R_i}|}, \tag{2.10}$$

is the Coulomb potential of nucleis and:

$$V_H[n(\vec{r})] = e^2 \int \frac{n(\vec{r'})}{|\vec{r} - \vec{r'}|} dr', \tag{2.11}$$

is called Hartree potential and represents the Coulomb repulsion by all electrons acting on the one-electron in the Kohn-Sham equations [10]. Furthermore, this part includes a Coulomb interaction between the electron and itself because all Kohn-Sham electrons are considered as part of the total electron density. The self-interaction contribution is unphysical, and its correction is one of many results that are collected together into the final potential in the Kohn-Sham Equation [2.12]. The V_{XC} potential includes the exchange and correlation contributions to the single-electron equations. This term can be described as a "functional derivative" of the exchange-correlation energy,

$$V_{XC}[n(\vec{r})] = \frac{\delta E_{XC}(\vec{r})}{\delta n(\vec{r})} \tag{2.12}$$

The electronic density $n(\vec{r})$ in Kohn-Sham approach is calculated from the Kohn-Sham wave function of the non-interacting system, as the sum of the squares of the eigenfunctions Φ_α of single-particle KS Hamiltonian [52]. Explicitly, the electron density is a sum of occupied states given by:

$$n(\vec{r}) = \sum_{\alpha,occ} |\Phi_\alpha(\vec{r})|^2 \tag{2.13}$$

The Hamiltonian for this Schrödinger-like equation, called the Kohn-Sham equation, gives

$$H_{KS}\Phi_\alpha(\vec{r}) = \varepsilon_\alpha \Phi_\alpha(\vec{r}) \tag{2.14}$$

By inserting KS Hamiltonian Eq.[2.9] in the Eq.[2.14],

$$[-\frac{1}{2}\nabla^2 + V_I(\vec{r}) + V_H[n(\vec{r})] + V_{XC}[n(\vec{r})]]\Phi_\alpha(\vec{r}) = \varepsilon_\alpha \Phi_\alpha(\vec{r}), \tag{2.15}$$

we can obtain Kohn-Sham eigenvalues and eigenfunctions (KS orbitals) by solving equation [2.15], taking the ground state density as given by equation [2.13].

2.2.5 Approximations to $E_{XC}[n]$

In DFT, Local Density Approximation (LDA) and Generalized Gradient Approximation(GGA) are the two most common approximations used for the exchange-correlation energy.

Local Density Approximation (LDA)

This is probably the most used approximation [10]. In brief, it is assumed that the charge density changes slowly, and each infinitesimal volume is treated like uniform electron gas. Then one obtains an integral form for E_{XC},

$$E_{XC}^{LDA}[n] = \int n(\vec{r})\varepsilon_{xc}(n)dr \tag{2.16}$$

Here ε_{xc} is the exchange-correlation energy per electron in a homogeneous electron gas with density n. This term can be calculated using, for example, Quantum-Monte Carlo or other accurate methods.

Generalized Gradient Approximation (GGA)

The Generalized Gradient Approximation (GGA) provides an enhancement in the accuracy compared to the one obtained by the LDA. In LDA, the ε_{xc} term depends on the value of the density at each point, but in GGA it also depends on the gradient of the electron density [48, 53],

$$E_{XC}^{GGA}[n] = \int n(\vec{r})\varepsilon_{xc}(n(\vec{r}), |\nabla n(\vec{r})|) \tag{2.17}$$

2.3 Self-Consistent-Field (SCF)

The Kohn-Sham equation [2.15] is an example of what is called a Self-Consistent Field (SCF) calculation [54], as is the Hartree-Fock (HF) method, for example. The fundamental form of any SCF algorithm requires iteratively changing some initial trial density matrix until a convergence point is reached, which means the output density is the same as, or sufficiently similar to, the input density [55]. To find the solution for the Kohn-Sham equations, we have to know the Hartree potential and to know the Hartree potential we have to know the electron density. However, to get the electron density, first, we need to know the single-electron wavefunction and to have these wavefunctions we should find solutions for the Kohn-Sham equations. Usually, this kind of problem can be solved iteratively. The first step is to make an initial guess for electron density $n(\vec{r})$. Then, plug it into the Kohn-Sham Hamiltonian and construct it based on estimated density Eq.[2.13]. The next step is to solve the Kohn-Sham equation using the estimated electron density to find the single-particle wavefunctions, Eq.[2.15]. Then, we construct a new electron density defined by the Kohn-Sham single-particle wavefunctions, take this new electron density and compare with the estimated electron density. If the initial density and the new density that we constructed are the same, then it can be considered as a converged ground-state electron density. However, if these two densities are not the same, then the estimated electron density should be changed to the new density that we obtained. This process should be repeated several times until we get the same densities, i.e., achieve convergence. The diagram [2.1] shows this iteration scheme.

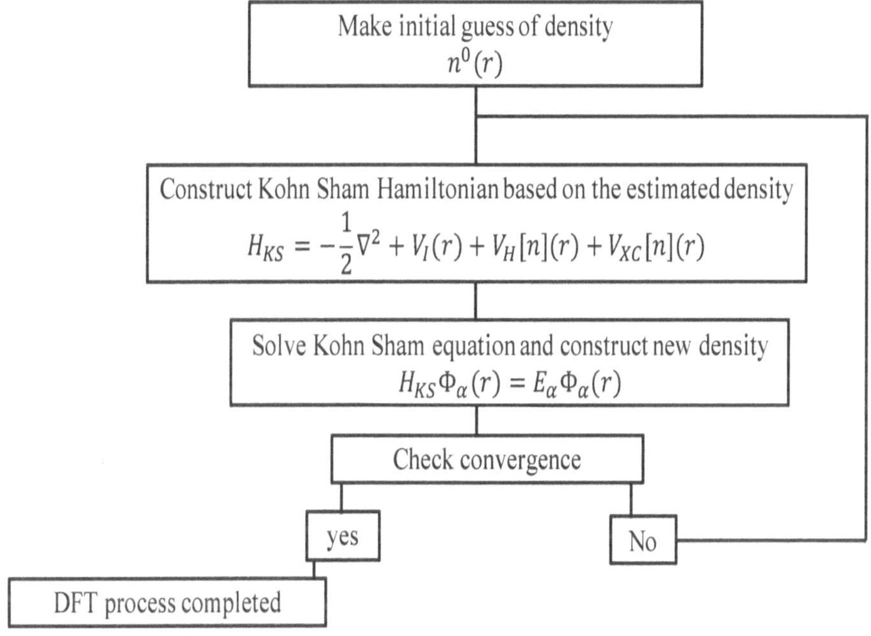

Figure 2.1: Density Functional Theory Iteration Scheme

2.3.1 Crystal Lattice Structure

The atoms in a crystal are arranged within unit cells, repeated in space. The lattice vector R is represented by:

$$R = n_1 a_1 + n_2 a_2 + n_3 a_3, \tag{2.18}$$

where n_1, n_2, n_3 are integer numbers, either positive or negative, and a_1, a_2, a_3 are the lattice (or primitive) vectors defining the three-dimensional unit cell with volume Ω. One unit cell can be occupied by one, two or a few number of atoms. The possible forms of a unit cell are limited by the fact that the periodic iteration of the unit cell must be space-filling, meaning no overlap or vacuum in the lattice. The lattices defined by equation [2.18] are called Bravais lattices [56, 57].

A crystal is a periodic arrangement of atoms that can be considered as infinite. In crystalline solids, a few numbers of atoms (a basis) is reproduced periodically over one, two, or three directions in space. There are numerous ways to characterize a crystalline solid, relying on the number of atoms in the unit cell that are periodically repeated. However, there are optimal ways that have all the symmetries of the system. Also, lattice vectors which mark the size of the unit cell, and the directions of the repetition are all important information to replicate the infinite crystalline structure. The lattice (or primitive) vectors are generally not orthogonal. The Wigner-Seitz cell is the one with the smallest volume possible, its limits are defined by planes localized in the half-distance between nearest-neighbor cells. Consequently, the classical simple cubic cell related to a bcc lattice is two times larger than the unit cell, and fcc is larger by four times.

2.3.2 Solving Kohn-Sham equation and Bloch's theorem

Now we have an expression for a Kohn-Sham equation for the non-intracting system given as:

$$[-\frac{1}{2}\nabla^2 + V_{KS}(\vec{r})]\psi_k^\alpha(r) = \varepsilon_k^\alpha \psi_k^\alpha(r),$$
(2.19)

where the V_{KS} potential is given by:

$$V_{KS}(\vec{r}) = V_I(\vec{r}) + V_H[n(\vec{r})] + V_{XC}[n(\vec{r})].$$
(2.20)

For the periodic system, the KS potential is a periodic potential. Since we are dealing with crystalline solids, this means that electrons are in this periodic potential.

The Bloch's theorems express the wavefunction of infinite crystal in terms of wave function at reciprocal space vectors of a Bravais Lattice [58]. We can write the wavefunction of the electron in an external periodic potential $v(r) = v(r + a)$ (where a is the length of the unit cell) as the product of a plane wave and a function with the matching periodicity of potential [57,59]

$$\psi_k^\alpha(r) = e^{ik.r} U_k^\alpha(r) \tag{2.21}$$

where

$$U_k^\alpha(r) = U_k^\alpha(r + a), \tag{2.22}$$

Here α is band index. So, the wavefunction at any point r can be replaced by wavefunction at $(r + a)$.

2.3.3 Plane-wave in DFT

The plane-wave approach, as we will show, is a useful way to study solid-state systems. The Bloch's theorem Eq.[2.21] illustrates the electronic wavefunction in the crystal which is a periodic environment [45]. The Bloch function $U_k(r)$, which brings the lattice potential periodicity, can be written by a discrete set of plane waves whose wave vectors are reciprocal lattice vectors (G) of the crystals as:

$$U_k^\alpha(r) = \sum_G C_G^\alpha(k) e^{iG.r}. \tag{2.23}$$

This summation runs over all reciprocal lattice vectors $G = m_1 b_1 + m_2 b_2 + m_3 b_3$. Thus, any function in a periodic system can be expanded according to Eq.[2.23]. Now, after combining Eq.[2.21] and Eq.[2.23] we will expand the wave functions in terms of liner combination of plane waves as:

$$\psi_k^\alpha(r) = e^{ik.r} \sum_G C_G^\alpha(k) e^{iGr} \tag{2.24}$$

where G is the reciprocal lattice wave vector, k is wave vectors and C_G are the expansion coefficients. Obviously, the Bloch's wavefunction can be represented as a sum of plane waves which have vector $(k + G)$ and the wavefunction of a periodic Hamiltonian can be expanded in a plane-wave basis set. The ground-state electron density of atoms is characterized by DFT. Kohn Sham potential energy can be expanded in a reciprocal space on the reciprocal lattice vectors G as:

$$V_{KS}(\vec{r}) = \sum_G V_G e^{iG.r} \tag{2.25}$$

Substituting this potential energy, Eq.[2.25], and periodic wave functions, Eq.[2.24], in Eq.[2.19] , the result is:

$$\left[-\frac{1}{2}\nabla^2 + \sum_G V_G e^{iG.r} \right] e^{ik.r} \sum_{G'} C_{G'}^\alpha(k) e^{iG'.r} = \varepsilon_k^\alpha e^{ik.r} \sum_{G'} C_{G'}^\alpha(k) e^{iG'.r}. \tag{2.26}$$

Thus:

$$\sum_{G'} -\frac{1}{2}\nabla^2 C_{G'}^\alpha(k) e^{i(k+G').r} + \sum_G \sum_{G'} V_G C_{G'}^\alpha(k) e^{i(G+k+G').r} = \varepsilon_k^\alpha e^{ik.r} \sum_{G'} C_{G'}^\alpha(k) e^{iG'.r}, \tag{2.27}$$

and ∇^2 acts on $e^{i(k+G').r}$, then we cancel $e^{ik.r}$ from all terms, and obtain:

$$\sum_{G'} \frac{(k+G')^2}{2} C_{G'}^\alpha(k) e^{iG'.r} + \sum_G \sum_{G'} V_G C_{G'}^\alpha(k) e^{i(G+G').r} = \varepsilon_k^\alpha \sum_{G'} C_{G'}^\alpha(k) e^{iG'.r}. \tag{2.28}$$

The second term has two sums over G and G'; therefore, we define $q = G' + G$ and change the sum over G' into sum of q then we relabel q as G. As a result, we get:

$$\sum_{G'} \frac{(k+G')^2}{2} C^{\alpha}_{G'}(k) e^{iG'.r} + \sum_{q} \sum_{G'} V_{(q-G')} C_{G'}(k) e^{i(q).r} = \varepsilon^{\alpha}_k \sum_{G'} C^{\alpha}_{G'}(k) e^{iG'.r} \qquad (2.29)$$

$$\sum_{G'} \frac{(k+G')^2}{2} C^{\alpha}_{G'}(k) e^{iG'.r} + \sum_{G'} \sum_{G} V_{(G-G')} C_{G'}(k) e^{i(G+G').r} = \varepsilon^{\alpha}_k \sum_{G'} C^{\alpha}_{G'}(k) e^{iG'.r} \qquad (2.30)$$

$$\frac{(k+G')^2}{2} C^{\alpha}_{G'}(k) + \sum_{G} V_{(G-G')} C^{\alpha}_{G}(k) = \varepsilon^{\alpha}_k C^{\alpha}_{G'}(k) \qquad (2.31)$$

Eq.[2.31] is the Kohn-Sham equation. This equation [2.31] is eigenvalue problem and it can be written in a similar way as $H\psi = E\psi$, where H are matrix elements, ψ are column vectors $C^{\alpha}_G(k)$ representing Kohn-Sham eigenfunctions, and ε^{α}_k are the energy eigenvalues. This matrix elements of H can be represented as, for the diagonal when $m = m'$:

$$H_{m,m'} = \frac{(k+G_m)^2}{2},$$

and for the off-diagonal matrix of H given when $m \neq m'$ as:

$$H_{m,m'} = \sum_{G} V_{(G_m - G_{m'})},$$

By diagonalizing this matrix, we can find the eigenvalues ε^{α}_k and corresponding eigenvectors $C^{\alpha}_G(k)$. The matrix size and the calculation accuracy depend on the number of reciprocal lattice vectors considered in our calculation. This is solved in a self-consistent way as the Kohn-Sham potential V_{KS} depends on the electronic density $n(\vec{r})$ which is the solution of Kohn-Sham equation, with a ground-state electron density defined as the following expression:

$$n^{\alpha}_k(\vec{r}) = \sum_{G} \sum_{G'} C^{\alpha}_{G'}(k) C^{\alpha}_{G}(k) e^{i(G'-G).r}, \qquad (2.32)$$

where the index α runs over the occupied states.

The cut-off energy shows the number of plane-wave functions used as basic functions to represent the wave functions. The more plane-waves we use in this expansion, the better wave- functions we get from this modeling. But, to solve this problem computationally, we should make a set of numerical approximations [45].

Let's explore some essential concepts in terms of the actual calculation with plane-waves in DFT. There are an infinite number of allowed G, but the coefficient C_G becomes smaller as G^2 becomes larger. The cut-off energy (E_{cut}) is defined by this expression:

$$E_{cut} = \frac{h^2}{2m}|G_{cut}|^2 \tag{2.33}$$

It is important to highlight that the lowest energy terms are more important (large weight) to the final solution than the high energy terms. Consequently, it is appropriate to reduce the infinite sum to just include solutions with kinetic energies smaller than some value, because, numerically we cannot deal with an infinite sum.

As a result, since we do DFT calculations, we have to introduce one parameter related to cut-off energy. The key point to remember here while defining the cut-off energy in our calculation is that, we do not consider plane-waves that have high kinetic energy. This is an essential input parameter whenever we do DFT calculation. Moreover, the way to determine an adequate cut-off energy is to do convergence tests in terms of the total energy to choose the minimal cut-off energy yielding accurate results. Figure [2.2] shows an example of the convergence of the total energy (E_{tot}) of a 2D graphene lattice as a function of the E_{cut} energy. At the final two steps the total energy has only a slight difference, of approximately

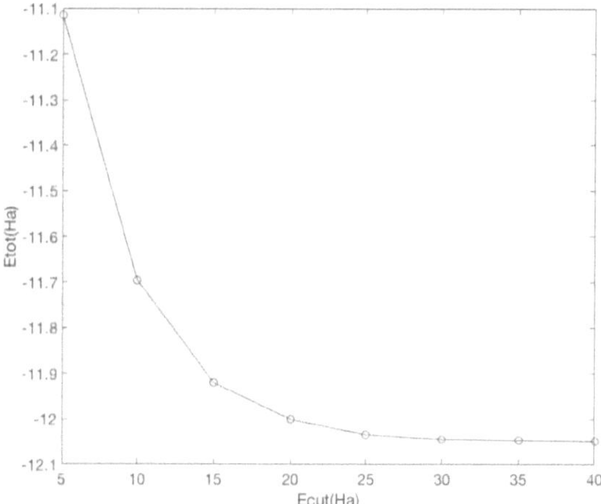

Figure 2.2: E$_{tot}$ as a function of the cut-off energy for monolayer graphene.

0.001Ha where 1Ha=27.211 eV [60].

K-points sampling

Sums and integrals over the Brillion Zone are important in DFT when we deal with insulators and semiconductors. For example, the calculation of the density for points of the BZ given as:

$$n(\vec{r}) = \sum \int_{BZ} |\psi_k(\vec{r})|^2 d^2k \tag{2.34}$$

Thus, for a periodic system this integral should be approximated by a sum over a discrete number of k-points. In this part, we will briefly introduce the criteria for choosing k-points,

30

which means determining the number of k-points that we have to take in our calculations in each direction in reciprocal space. This is important if we have periodic directions in our simulated cell, and also, to reduce the computational cost of infinite integrations in DFT. The total energy of any system in DFT is obtained by summation over all its electrons [57]. There are many studies that try to keep this problem under control. From the previous introduction about numerical integration, it is obvious that choosing a good (big enough) number will lead to accurate results. However, one still needs to determine the number of k-points needed for the calculations [45]. The number of k-points should be large enough to obtain reliable energies and get total energy convergence. The Table [2.1] shows results for some calculations done by DFT in Abinit for 2D graphene structure [61]. Each value in this table illustrates the total energy as a function of the number of k-points in the Brillouin zone in x and y directions, since for 2D materials the number of k-points in z direction is typically one. This calculation used a hexagonal unit cell with two atoms per unit cell. The results in Table [2.1] are plotted in Figure [2.3]. By comparing the energies in the table (or the graph) it is clear that when the number of k-points are greater or equal to around 14, the total energy is converged well enough and changes only in the order of 0.001 eV.

Table 2.1: Number of K-points (x and y directions) and total energy.

number of k-points	4	7	10	14	19	24	30	37
Total energy(Ha)	-12.048	-12.046	-12.047	-12.047	-12.047	-12.047	-12.047	-12.047

Indeed, convergence is an important concept that we will repeat many times. As we perform DFT calculations, we have to check whether the calculations are converged.

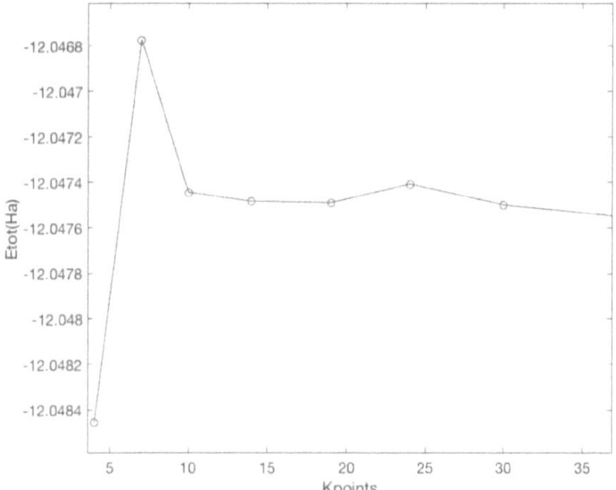

Figure 2.3: Total energies for monolayer graphene calculated as described in table [2.1] as a function of number of K points

So, for any system, we have to figure out two convergence parameters, the Brillouin zone sampling and the cut-off energy in reciprocal space. These parameters (cut off energy and k-points grid) are important in our calculation due to the difficulty in dealing with infinite summations [44].

2.3.4 Pseudopotentials

Due to the strong ionic potentials and experimentally highly localized states of core electrons, it is challenging to deal with this region. Most of the recent electronic structure calculations tend to replace the nuclear attractive potentials and the atomic core electrons potentials by a smooth potential, creating slowly varying densities in the core of the atoms. In this approach, valence states and the core states should be orthogonal. Therefore, instead of standard electron-ion interaction, the smooth pseudopotential will be used in the calculations. Also, pseudoatomic wavefunctions require a smaller number of plane-waves because potential oscillates less. Generally, the electrons of the core behave as frozen electrons, so they do not contribute to the chemical bonds. Thus, the only contribution comes from valence states. Many important projects are designed to use the pseudopotential approximation.

Chapter 3

Results and discussions

In this chapter, we present and discuss results obtained by the application of DFT to different structures. We start with graphene, h-BN and MoS2 monolayers, next we proceed to graphene and h-BN bilayers, and finalize with the study of heterostructures h-BN/graphene for different configurations. Our calculations have been performed using the Abinit package [61]. Abinit's main program is based on Density Functional Theory (DFT). The total energy, ground state ,density, and electronic structure for a given material can be calculated by Abinit. It uses pseudopotential to describe the chemical bonding and it uses plane waves basis to describe the Kohn Sham single particle orbital [61]. .

3.1 Electronic structure of graphene, h-BN, and MoS$_2$ monolayers

In this section, we study electronic properties of graphene, MoS$_2$, and h-BN monolayers. Calculations were performed in the Density Functional Theory (DFT) framework, with a plane-wave basis set, and the pseudopotential method.

Table 3.1: Unit cell parameters a and c, total energy E_T in Hartree per unit cell, and gap energy E_g in eV for graphene, h-BN, and MoS$_2$ monolayers.

Structure	a(Å)	c(Å)	E_T(Ha)	E_g(eV)
graphene	2.45	9.50	-12.05	0.00
h-BN	2.52	6.70	-13.41	4.68
MoS$_2$	3.12	15.14	-89.90	1.67

3.1.1 Graphene

Calculations, in the Abinit code [61], were performed in the Generalized Gradient Approximation (GGA) for the exchange-correlation term [62], with norm-conserving pseudopotentials. Lattice parameters, obtained by the relaxation of the graphene lattice, are presented in Table [3.1]. The cut-off energy was set to 680 eV, and the k-grid, with a Monkhorst-pack mesh, was limited to [8 × 8 × 1] mesh, to reduce the computational cost. This set of parameters led to fast calculations and well-converged results, as mentioned in Chapter 2. Figure [3.1]

shows the honeycomb lattice of graphene. $\vec{a}_1$ and $\vec{a}_2$ are the lattice vectors given by:

$$\vec{a}_1 = a(1,0), \tag{3.1}$$

$$\vec{a}_2 = a\left(-\frac{1}{2}, \frac{\sqrt{3}}{2}\right), \tag{3.2}$$

where $a = 2.45$ Å is the unit cell parameter. Each unit cell contains two atoms belonging

to two different sublattices (sublattice A and sublattice B) where A sublattice atom is

surrounded by 3 B sublattice atoms and vice versa.

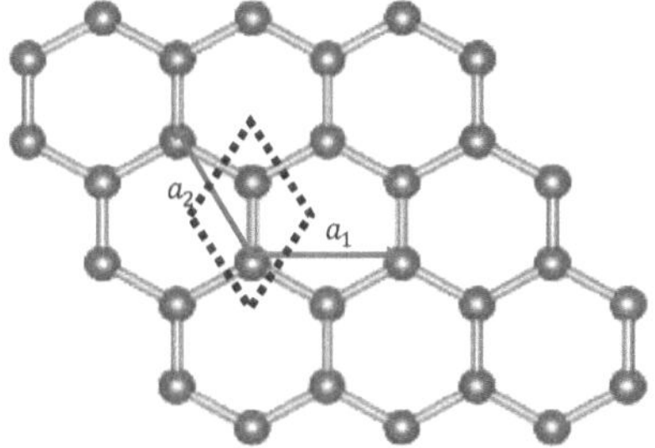

Figure 3.1: Graphene honeycomb lattice structure. The red arrows are the basis vectors $\vec{a}_1$

and $\vec{a}_2$ of the Bravais lattice and the unit cell is indicated by the dashed lines.

To get a 2D structure of a monolayer, we increase vertical separation c between the

layers to prevent their interaction. This interlayer distance $c = 9.50$ Å is presented in the

Table [3.1] as the c parameter. For the graphene relaxed unit cell, the calculated a is 2.45Å.

The obtained band structure is shown in Figure [3.2a]. The band structure of a single layer shows a linear dispersion, associated with the propagation of massless Dirac fermions, around the Fermi level E_F. At the six K-points in the Brillouin zone, the valence band maximum touches the conduction band minimum, producing zero bandgap. These touching points happen between bonding π and anti-bonding π^* orbitals, at each valley K and K', implying vanishing hopping between nearest neighbours, or sub-lattices. At these special points, the electronic dispersion is linear, corresponding to electrons with zero effective mass. These electrons behave as relativistic massless Dirac particles. At low energy, for small deviations (q) around the K-point, the effective Hamiltonian is,

$$H_K = v_F \begin{pmatrix} 0 & (q_x - iq_y) \\ (q_x + iq_y) & 0 \end{pmatrix}, \tag{3.3}$$

with eigenvalues given by:

$$E = \pm v_F \mathbf{q}, \tag{3.4}$$

where $v_F \simeq 10^6 m/s$ here is the Fermi velocity which is analogues to that of photons. The maximum electron speed in graphene is $v_F/c = 1/300$, according to [63], compared with the speed of light in vacuum [10]. In terms of the Pauli matrices σ, the Hamiltonian, Eq.[3.3] , reads

$$H_K = v_F \vec{\sigma}.\vec{q}. \tag{3.5}$$

Eq.[3.5] illustrates that the electron behaves as a massless relativistic particle around the K-point . Figure [3.2b] shows a zoom on the DFT energy bands around the K-point confirming the linear dispersion. The Fermi velocity v_F obtained for graphene suspended in air from Figure 3.2b is $v_F = 0.85 \times 10^6 m/s$ which is proportional to the speed of light.

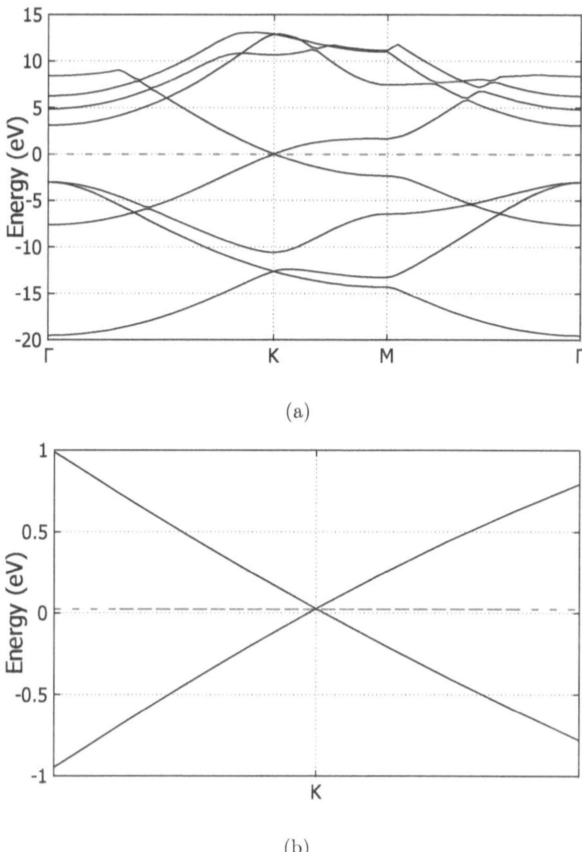

(a)

(b)

Figure 3.2: (a) The ab-initio band structure of Graphene. (b) The conduction band and valence band touching each other at the K point.

Figure [3.3] shows the calculated electronic charge density of the graphene structure. In this plot, we can observe small purple areas with no electronic charge density around the atomic positions, which refers to low density in this area due to P_z orbital. Also, the yellow

38

areas indicate high charge density. This effect comes from the sp^2 hybridization in carbon atoms.

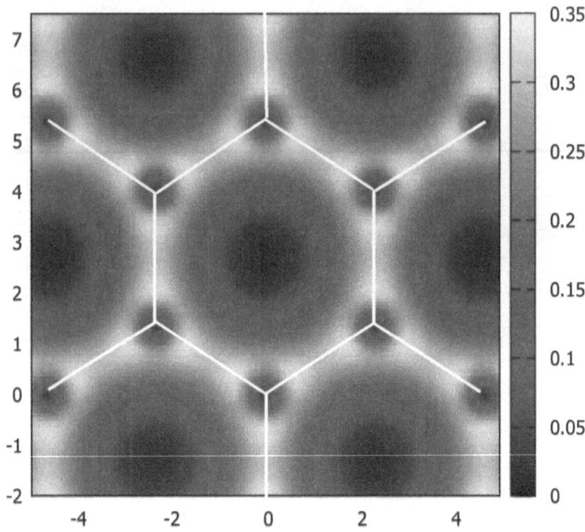

Figure 3.3: Electronic density of graphene on the z=0 plane.

3.1.2 Single layer of MoS$_2$

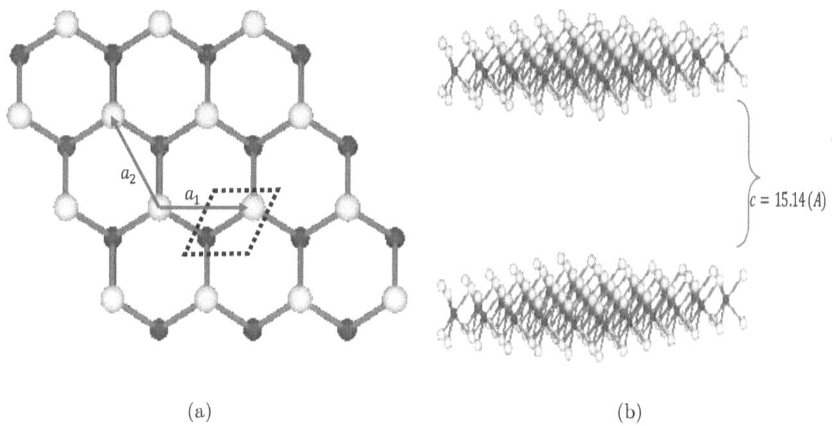

Figure 3.4: (a) MoS$_2$ crystal structure of a hexagonal lattice of Mo (black) atoms and S (yellow) atoms with unit cell indicated by dashed line of 3 atoms (1Mo, 2S). (b) side view shows three planes in this order S-Mo-S. "c" refers to the distance between MoS$_2$ layers.

MoS$_2$ structure is shown in Figures [3.4a] and [3.4b]. In this structure we have 3 atoms per unit cell (1 Mo, 2 S). Lattice vectors are given by $\vec{a}_1$ and $\vec{a}_2$ (3.1 and 3.2 expressions) and $a = 3.12$ Å is the unit cell parameter. Distance between Mo atom and S up and down is 1.56 Å. To get a single layer structure, we separate layers by 15.14 Å between layers to prevent interlayer interaction. The MoS$_2$ electronic band structure was calculated using the DFT-LDA approach, results are shown in Figure [3.5]. In this figure, we present results without spin-orbit coupling [3.5a] SOC and with SOC [3.5c]. In Figure [3.5a] we observe that the energy gap opens and the bandgap along the high symmetry path (ΓKMΓ) of MoS$_2$ monolayer, without SOC, is direct. The bandgap value obtained in our calculation was

$E_g = 1.67$, eV Figure [3.5b], in reasonable agreement with the experimental value of ≈ 1.8 eV [30]. In a similar procedure as the one applied to graphene monolayer, we can explore a Dirac-like Hamiltonian around the K-point in MoS_2. This Hamiltonian presents Massive Dirac Fermions, introduced by the opening of a gap δ,

$$H_K = \begin{pmatrix} \frac{\delta}{2} & v_F(q_x - iq_y) \\ v_F(q_x + iq_y) & -\frac{\delta}{2} \end{pmatrix}, \tag{3.6}$$

$$H_K = v_F \vec{q}\vec{\sigma} + \frac{\delta}{2}\vec{\sigma}_z. \tag{3.7}$$

From our DFT calculation we can infer the δ parameter as the gap opening in Figure [3.5b] at K-point, $\delta = 1.68$ eV. Another characteristic of MoS_2 is the strong spin orbit coupling (SOC) which causes the splitting between bands with different spin. Complementing this analysis, Figure [3.5c] shows the electronic band structure of MoS_2 including SOC correction. It can be seen, from Figure [3.5d], that it is still a direct bandgap semiconductor. However, there is a splitting in the valence band at K-point of about $\simeq 150$ meV due to the spin-orbit coupling [9, 64]. The presence of spin orbit coupling is exactly what makes this material interesting and suitable for application in nanodevices [9, 65] .

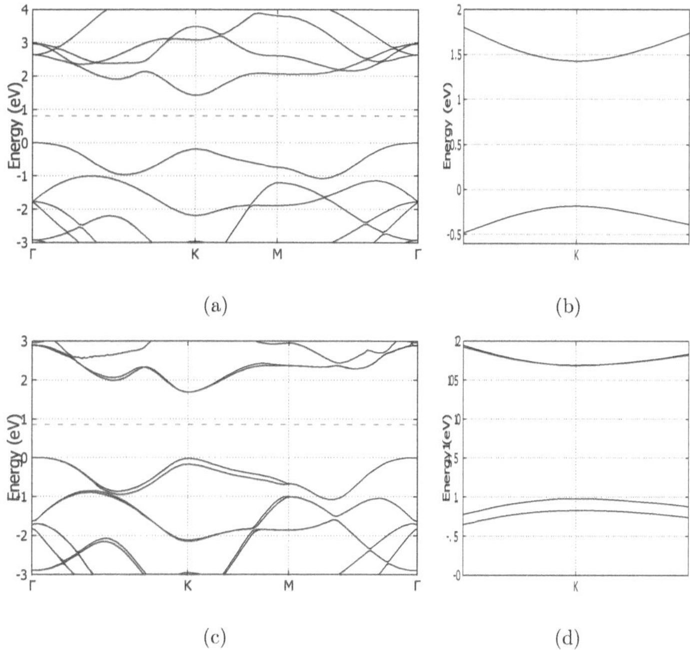

(a) (b)

(c) (d)

Figure 3.5: Band structure of a monolayer MoS$_2$: (a-b) No SOC, bands along the ΓKMΓ and zoom around the band gap at K point, respectively; (c-d) Same as (a-b), but including SOC corrections.

Moreover, the conduction band has another minima besides the K-points. This local minima (Q) are localized along ΓK paths, with a difference in energy of 0.3 eV in comparison with the global minima in the K-point. Figure [3.6a] shows these two minima points [9]. These two minima play an important role in determining the electronic properties of this material.

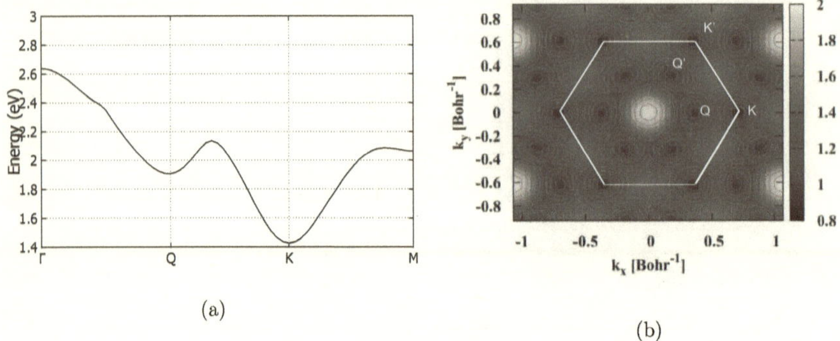

Figure 3.6: Bottom of conduction band: (a) Following ΓKM path, where we can see the two minima points, Q and K. (b) In the Brillouin zone, with minima Q and K, and the non-equivalent Q' and K' marked [9].

3.1.3 h-BN monolayer

Similar to graphene, h-BN has a hexagonal structure; however, instead of carbon atoms, boron (B) and nitrogen (N) occupy its two sub-lattices as shown in Figure [3.7]. We insert a big vacuum area between the layers to prevent their interaction. This interlayer distance, $c = 6.70$ Å, is presented in the Table 3.1 as the c parameter.

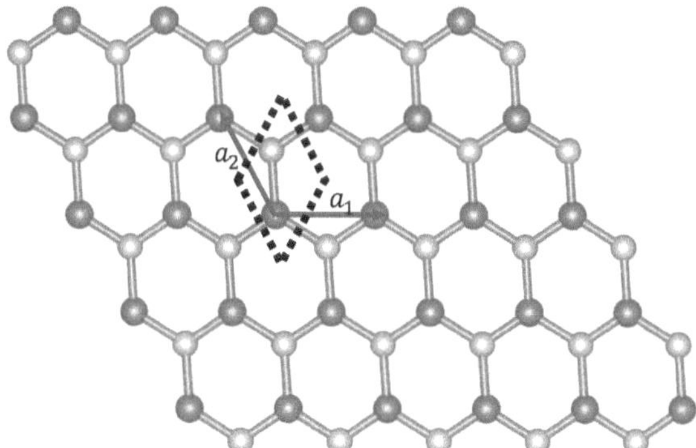

Figure 3.7: h-BN honeycomb lattice structure (boron (B)=green,Nitrogen (N)=blue). The red arrows are the basis vectors $\vec{a}_1$ and $\vec{a}_2$ of the Bravais lattice and the unit cell is indicated by the dashed area.

We obtain the electronic properties, using the same values as in graphene for cut-off energy (680 eV) , k-grid $[8 \times 8 \times 1]$, and optimized unit cell parameter, $a = 2.52$ Å, presented in the Table [3.1]. To perform the calculations, we made use of norm-conserving pseudo-potentials within the GGA approach [62]. In Figure [3.8a], we show the band structure of h-BN where we observe it is an insulator with a large direct bandgap, $E_g = 4.68$ eV, in K-point as expected [66]. The bond between B and N atoms is created from the combination of B and N sp^2 orbitals. However, because of the difference in electronegativity between B and N atoms, electrons are transferred from B to N. Consequently, in contrast with purely covalent bonds in graphene, the bonding between B and N gains an ionic character.

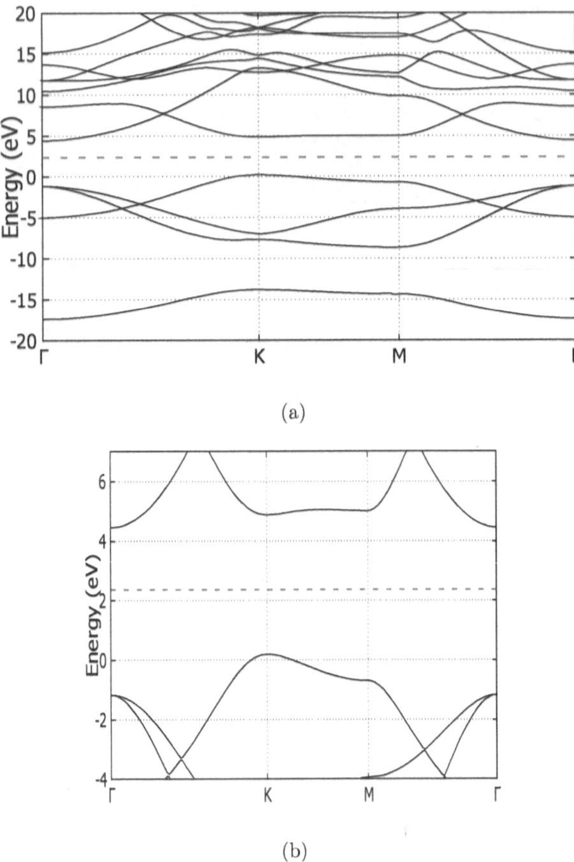

Figure 3.8: (a) Band structure of h-BN monolayer. (b) Same as (a), but zooming in the band gap at K point.

The charge transfer from B to N influences the properties of h-BN, for example, opening a large bandgap [67]. Figure [3.9] shows the charge imbalance between B and N species. The peak in charge density is localized around N atoms which appears in the image as a bright colour, in opposition to the weak electron localization around B atoms. The most important feature of this 2D material in comparison with other 2D material is the unique lattice constant that is very similar to graphene, and a large bandgap.

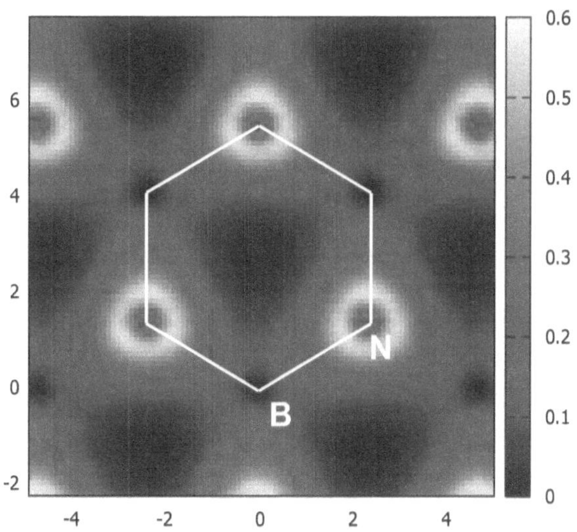

Figure 3.9: Electronic density of h-BN on the z=0 plane.

3.2 Electronic structure of bilayers of graphene and of h-BN

In this section, we investigate the electronic properties of bilayers of graphene and h-BN. In addition, we perform comparisons with their respective monolayers, presented in the previous section. All calculations were performed in the Abinit package. For the exchange-correlation,

46

Local Density Approximation (LDA) was applied, and norm-conserving pseudopotentials were used. Also, for both structures, the cut-off energy was set to 680 eV and the k-grid to [10 × 10 × 1] with a Monkhorst-pack mesh. In all cases, the unit cell was relaxed with optimized cell parameters, including inter-layer distance, presented in the Table [3.2].

Table 3.2: Optimized cell parameters a and interlayer-distance d, total energy (E_T) and bandgap (E_g) for bilayers graphene and h-BN.

Structure	a(Å)	d(Å)	E_T(Ha)	E_g(eV)
graphene	2.46	3.36	-24.24	0.00
h-BN	2.50	3.36	-24.12	4.35

3.2.1 Bilayer graphene

We first consider bilayer graphene with two layers on top of each other with AB stacking and a unit cell consisting of 4 carbon atoms as in Figure [3.10].

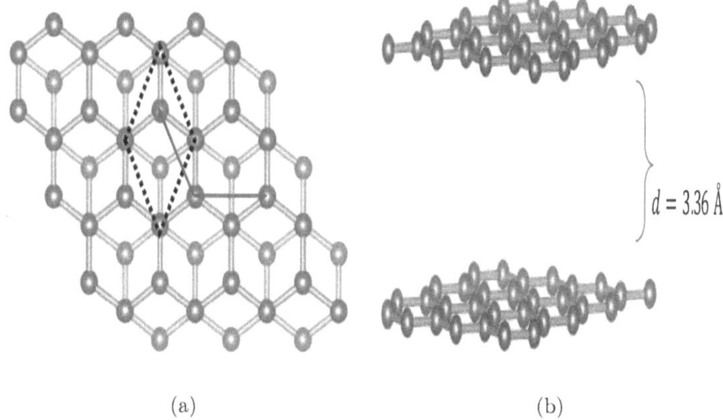

(a)   (b)

Figure 3.10: (a) Graphene bilayer with AB-stacking. The unit cell of bilayer graphene consists of 4 atoms in each unit cell, (b) side view of bilayer graphene, where d is the distance between these two layers.

The Figure [3.11a] shows the obtained band structure. Our results are consistent with what is expected from experiments and theoretical calculations [68, 69]. Similar to monolayer graphene, bilayer graphene is a semiconductor with zero energy gap. However, we have two pairs of parabolic bands (See Figure 3.11b). The two bands touch at the K-point and the two π-bands are split by ≈ 0.8 eV. This is an example of massive gapless Dirac fermions [10].

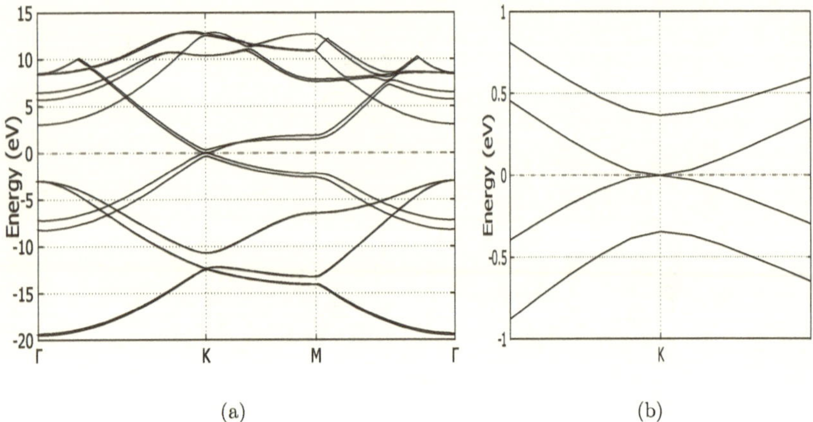

(a)   (b)

Figure 3.11: (a) Band structure of bilayer graphene. (b) Parabolic band structure of bilayer graphene near the K point.

Unlike graphene, the bilayer graphene shows a parabolic dispersion around the K-point, as can be seen in our DFT calculations (See Figure 3.11b). This parabolic dispersion comes without the opening of a gap [70]. In AB stacking, the low energy Hamiltonian can be written as:

$$
H_{BG} = \begin{pmatrix}
0 & v_F(q_x - iq_y) & 0 & 0 \\
v_F(q_x + iq_y) & 0 & \gamma & 0 \\
0 & \gamma & 0 & v_F(q_x - iq_y) \\
0 & 0 & v_F(q_x + iq_y) & 0
\end{pmatrix} \tag{3.8}
$$

where γ is the nearest interlayer neighbor coupling, which relates the splitting between the π and $\pi*$ bands. By our calculations we estimate $\gamma = 0.4$ eV. For the two touching bands

49

the Hamiltonian in the of K-point can be approximated as:

$$H_{BG}^{u} = -\frac{v_F^2}{\gamma} \begin{pmatrix} 0 & (q_x - iq_y)^2 \\ (q_x + iq_y)^2 & 0 \end{pmatrix}. \tag{3.9}$$

This Hamiltonian describes the coupled bands around the K-point. Bilayer graphene gives us a parabolic dispersion with zero bandgap (See Figure 3.11b). From DFT calculation, we obtain $\gamma \simeq 0.4$ eV and effective mass $m^* \simeq 0.0352\, m_e$ in reasonable agreement with [71–73].

3.2.2   Bilayer h-BN

The bilayer h-BN structure consists of two layers with AB stacking, with N atoms of layer 1 on top of B atoms of layer 2, and B atoms of layer 1 on top of the center of the hexagonal ring in layer 2, Figure [3.12].

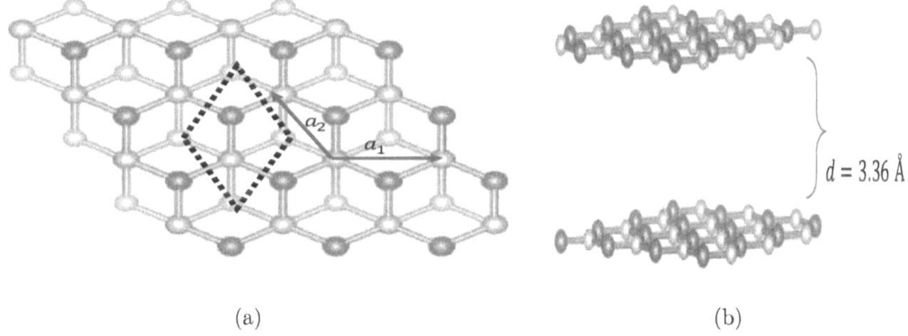

Figure 3.12: (a) AB-stacked bilayer and the unit cell of bilayer h-BN consists of 4 atoms in each unit cell, (b) side view of bilayer h-BN. d is the distance between these two layers.

From Figure [3.13a] we can notice that the bilayer h-BN has an indirect bandgap which contrasts with its monolayer counterpart. Its indirect bandgap is between a point close to K (valence band) and M (conduction band), as can be observed in Figure [3.13b] $E_g = 4.35$ eV in reasonable agreement with [67]. Also, the bandgap decreased by 0.33 eV in comparison with h-BN monolayer. The hybridization between the bands in its constituent monolayers shifts slightly the positions of maximum and minimum in the valence and conduction bands, respectively. In particular, the minimum of the conduction band is displaced to around the M point.

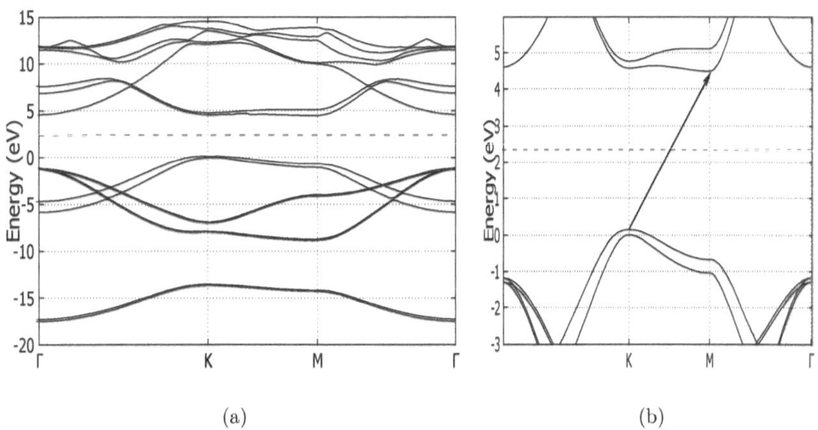

(a)   (b)

Figure 3.13: Band structure of bilayer h-BN.

The bandgap, now indirect, is gradually decreasing by increasing the number of layers.

3.3 Electronic structure of graphene and h-BN heterostructure

In this section, we study the electronic properties of heterostructure, consisting of h-BN layer on top of a graphene layer and explore different layer configurations. After this we find the most stable configuration and discuss important properties of these configurations such as the bandgap.

This supercell of heterostructure is created by a 1:1 ratio between h-BN and graphene unit cells. There is a small lattice mismatch between these two layers, less than 2% [74]. In this study, we consider three different types of configurations. Because of using 4 atoms per unit cell , Moiré superlattice does not exist in our calculations which it needs a large number of atoms in the unit cell. Regarding the stacking configuration, we show that this is critical aspect for determining the properties of heterostructures [75]. The first is the AA stacking, in which the C atoms are directly below the B and N atoms. The second configuration is in the AB stacking (Bernal stacking), with the first C atom below the N atom [C-N] and second C atom below the B atom [C-B]. In the third configuration the second C atom is below the hexagonal ring (see Figure 3.14). In our calculation, we did not consider twisting among layers of graphene and h-BN layers.

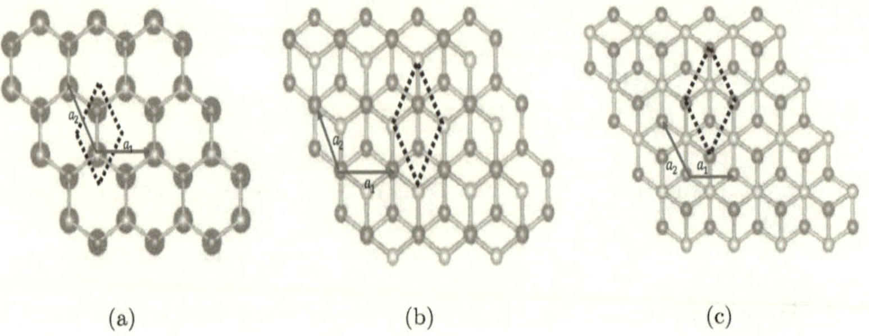

(a)   (b)   (c)

Figure 3.14: (a) Different configurations of C, B and N atoms in graphene/h-BN structure with their unit cell : (a) (AA), (b) (AB[C-B]), and (c) (AB[C-N]).

These different configurations are shown in Figure [3.14].

This calculation was performed within the local density approximation (LDA) for the exchange-correlation term, using norm-conserving pseudo-potentials. For all configurations, cut-off energy was set to 680 eV and the k-grid to $[10 \times 10 \times 1]$ with a Monkhorst-pack sampling. Also, in-plane lattice parameters were obtained by cell optimization and are presented in Figure [3.15a] and interlayer distance in Figure [3.15b] for the three different configurations. Interlayer distance varies among different configurations, with values of 3.47, 3.45, and 3.35 Å for (AA), (AB)[C-N], and (AB)[C-B] configurations, respectively.

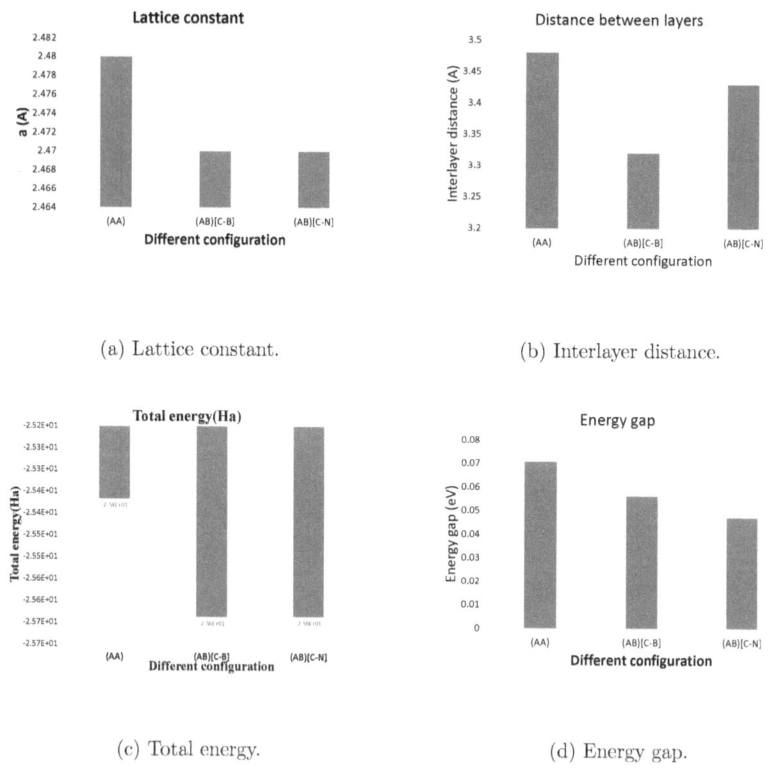

(a) Lattice constant.

(b) Interlayer distance.

(c) Total energy.

(d) Energy gap.

Figure 3.15: Physical quantities for different configurations. Here AA label refers to AA stacking and AB label to AB stacking with [C-B] and [C-N] labels specifying which atoms are above the C atom.

Furthermore, from the calculation, for all three configurations there is an open bandgap in graphene with different values. Also, we found a more stable configuration between them which is the (AB) [C-B] structure which has a lower total energy in comparison with other configurations. The total energy of the (AB)[C-N] structure increased by approximately 0.0007 (Ha). The (AA) Stacking has less stability than either (AB) stacking configurations. Qualitatively, this is due to interaction among the electrons in graphene and the cations in BN and the repulsion between the electrons in graphene and the anions in BN. The N atoms tend to be above the center of graphene hexagon where the electron density is lower. In contrast, B tends to be just above the C atom, maximizing the attractive potential felt by the electrons, and consequently, minimizing the total energy.

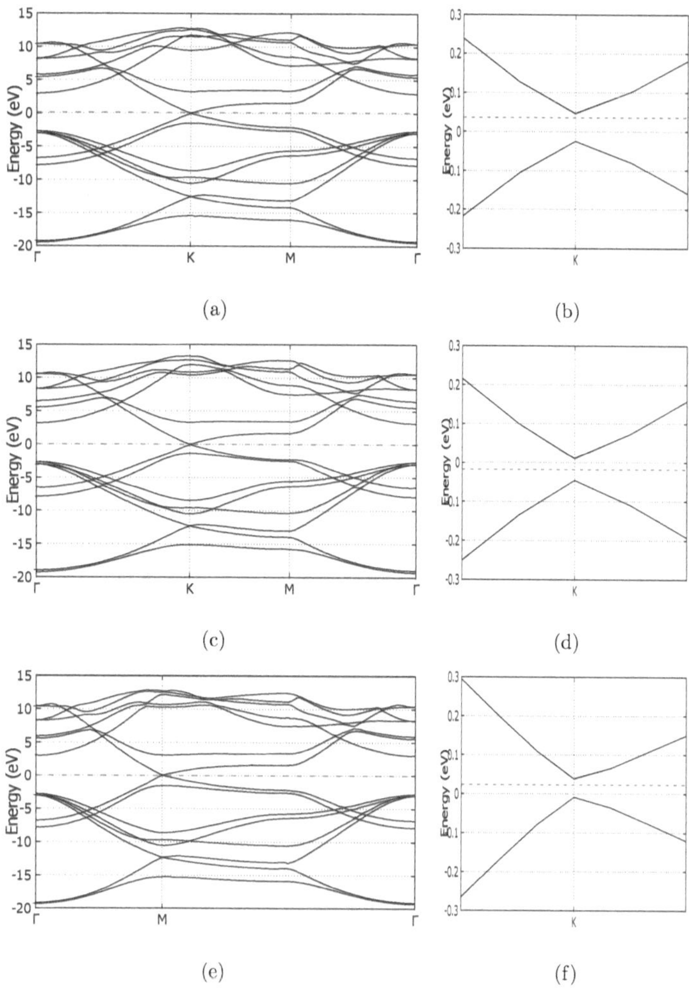

Figure 3.16: Band structure of graphene/h-BN heterostructures for different configurations where the (a-b) refer to AA stacking, (c-d) to (AB)[C-B], and (e-f) to (AB)[C-N].

From Figure [3.16] we can observe that all three configurations have almost the same band structure and all of them affect graphene band structure by opening a gap at K-point. The difference among them is in the value of the bandgap in graphene. AB[C-N] has a lower value of gap by about ≈ 0.05 eV compared to AB[C-B] [76].

Concerning the electronic structure properties for these configurations we can see from Figure [3.15d] that (AA) stacking has the maximum bandgap opening of 0.07 eV, whereas the minimum gap is obtained in the configuration (AB)[C-N] ≈ 0.05 eV [77].

3.4   Conclusion

In this work, we used the Density Functional Theory to determine the electronic structure of important monolayers and bilayers in the field of 2D materials. Studying the electronic structure has played an essential role in understanding the physical properties of these materials.

Our study considered different structures of monolayers graphene, MoS_2, and h-BN. We observed massless Dirac Hamiltonian in graphene with a zero bandgap. For $MoS_2$2, we found massive gapped Dirac fermions around the K-point; also, we show a strong splitting of its bands due to SOC.

For h-BN, a significant ≈ 5 eV direct bandgap was observed at the K-point. This material is of extreme importance as an insulator with lattice matching with graphene. This study also looked at graphene and h-BN bilayers and compared the results with its monolayers counterparts. Results show that bilayer graphene is still a gapless material. We observed massive Dirac fermions, due to the interaction between its two layers. h-BN bilayer, in contrast with its monolayer, has an indirect bandgap; also, the bandgap decreased by about 0.33 eV in comparison with monolayer h-BN.

Our primary objective was to study the heterostructure of h-BN and graphene. We observed a gap opening in the graphene band for all three h-BN /graphene configurations. The findings confirmed that the most stable structure is the AB stacking with C atom under B atom. Our calculations do not allow for twisting of h-BN-graphene layers and the appearance of Moiré pattern. Recently, h-BN has gained some attention as an insulating substrate with small lattice mismatch for graphene.